Liaison Functions

Held at the 2024 AIChE Annual Meeting

San Diego, California, USA
27-31 October 2024

ISBN: 979-8-3313-1688-4

Liaison Functions

Held at the 2024 AIChE Annual Meeting

San Diego, California, USA
27-31 October 2024

AIChE
The Global Home of Chemical Engineers

Printed from e-media with permission by:

Curran Associates, Inc.
57 Morehouse Lane
Red Hook, NY 12571

Some format issues inherent in the e-media version may also appear in this print version.

Printed with permission by Curran Associates, Inc. (2025)

For permission requests, please contact AIChE
at the address below.

AIChE
120 Wall Street, FL 23
New York, NY 10005-4020

Phone: (800) 242-4363
Fax: (203) 775-5177

www.aiche.org

Additional copies of this publication are available from:

Curran Associates, Inc.
57 Morehouse Lane
Red Hook, NY 12571 USA
Phone: 845-758-0400
Fax: 845-758-2633
Email: curran@proceedings.com
Web: www.proceedings.com

TABLE OF CONTENTS

EMERGING JUNIOR INVESTIGATOR OPEN INNOVATION FORUM (INVITED TALKS)

KICHE-US CHAPTER OPEN FORUM (INVITED TALKS)

ANDREAS ACRIVOS AWARD FOR PROFESSIONAL PROGRESS IN CHEMICAL ENGINEERING LECTURE

CATO T. LAURENCIN REGENERATIVE ENGINEERING FOUNDER'S AWARD LECTURE

IACCHE'S JAMES Y. OLDSHUE LECTURE

JOHN M. PRAUSNITZ AICHE INSTITUTE LECTURE

WILLIAM R. SCHOWALTER LECTURE

AICHE JOURNAL FUTURES: NEW DIRECTIONS IN CHEMICAL ENGINEERING RESEARCH (INVITED TALKS)

538a Point-of-Clinic Sensor for Precision Medicine to Tackle Diabetic Retinopathy

Sangmoo Jeong

About 10 million people in the US suffer from diabetic retinopathy, which can cause severe vision impairment or even blindness if not managed properly. These conditions are closely related to abnormal blood vessel growth in the retina, and therefore, the first-line treatment is the intravitreal injection of angiogenesis inhibitors, such as antibodies targeting VEGF and/or Ang2. However, no method currently exists in a clinic to determine which inhibitors would be more effective for each patient. Inaccurate or delayed treatments can result in permanent visual loss. Effective technologies are urgently needed to identify which molecules need to be targeted for each patient in a clinic. To address this critical problem, we have developed a rapid and sensitive diagnostic technology that can quantify cytokine levels in aqueous humor in a local clinic. Termed Immuno-Magnetic Oximeter-based Sensor (IMOS), it can quantify multiple cytokines as low as 25 pg/mL in <50 μL aqueous humor within 30 min. Moreover, it has the form factor of a smartphone. Its detection power comes from an oximeter module, which is widely used in a portable fingertip pulse oximeter device. We are the first to use its advantages for biomolecular detection. Our technology has the potential for precision medicine to tackle diabetic retinopathy. The technology details will be presented in this talk.

538b Modular Protein Self-Assembly: Building Bio-Inspired Materials with Genetically Programable Functionality

Won Min Park

Proteins are the essential biological building blocks in living systems to build sophisticated biological assemblies with highly controllable properties and functions. Significant effort has been devoted to mimicking the processes of protein self-assembly toward advanced materials, but it often fails to replicate the complexity and efficiency of native systems. We have found that by using protein zippers we can create modular systems to genetically program functions into self-assembled protein biomaterials. Specifically, coiled coils, the α-helical protein motifs that oligomerize with high affinity and specificity have been widely used as versatile toolkits to create functional protein assemblies. The sequence-to-structure relationships of coiled coils have been understood well, leading to the de novo design of coiled-coil sequences with controllable specificity, oligomeric state, length, and helix orientation. Exploiting the coiled-coil interactions through genetic fusion of recombinant proteins, we have demonstrated modular protein self-assembly to create functional protein biomaterials. Using this approach, we have designed fluorescent protein nanoclusters to image multiplexed signaling network dynamics in cells. Additionally, we have engineered biocompatible fluorescent coating materials that quantify calcium ions in biological systems to detect abnormal physiological conditions such as mild or moderate hypercalcemia. Finally, by combining a natural coat protein with a coiled-coil protein motif, we have developed a modularly programmable platform to build antifouling and antimicrobial surfaces. Through these technologies, we have consistently demonstrated that the genetic combination of coiled-coil toolkits and functional protein domains enables highly modular functional biomaterials engineering that could be extended to a variety of applications.

538c Radiation-Induced Modifications of Zeolitic Imidazolate Frameworks Membrane Characteristics

Dennis Lee

Zeolitic imidazolate frameworks (ZIFs) display remarkable flexibility in both their chemical composition and structure. Particularly intriguing are post-synthetic modification (PSM) methods, which offer the ability to precisely adjust MOF structures, enhancing their effectiveness in various applications such as gas storage, molecular separations, and catalysis.

In this presentation, we will introduce the solvent-less PSM approaches we employed, such as electron beam irradiation and nonthermal plasma treatment. These methods have been proven to enhance the CO_2 selectivity of ZIF-8 membranes over other gas molecules within a short treatment time of less than 2 minutes, at the expense of CO_2 permeation. Furthermore, we will describe various ex situ techniques (e.g., X-ray photoelectron spectroscopy (XPS), grazing incidence X-ray diffraction (GIXD), and electron microscopy) as well as in situ technique of infrared reflection absorption spectroscopy (IRRAS) used in collaboration with Brookhaven National Lab to provide insights into the modifications in the materials surface chemistry and their correlation with the altered membrane properties.

Moreover, we will discuss the recently obtained results of tuning the permselectivity of ZIF-8 membranes while leaving CO_2 permeance almost unaffected through X-ray exposure. Alongside materials characterizations and analysis based on the aforementioned techniques, we estimated the improvement in membrane separation performance resulting from X-ray irradiation. This estimation was done by calculating the separation index ratio (SIR), which is the ratio of the separation index (SI) after X-ray treatment to the value determined for the pristine membrane.

538d Gelation Behavior of Thermoreversible Diels–Alder Networks and Their Application for Reversible Adhesives

Youngmin Lee

Thermoreversible networks using the Diels–Alder (DA) chemistry can switch between a network state and a segment state. When they are introduced into a thermoset polymer such as epoxies, the network state provides advantages of thermoset polymers, for example, strong mechanical properties and excellent resistance against chemicals and environmental factors. At an elevated temperature (~120°C), however, it turns into a segment state through the retro-DA reaction, which permits self-healing, recycle, and 3D printing processes. In this talk, gelation behavior of DA epoxies and their application for reversible adhesives will be presented. Study of network formation from precursors is an important fundamental topic, but it also has significance in the application aspect. The current 3D printing approaches for thermoset polymers require a long post-curing step (e.g., 6 – 48 hours) to reach the full mechanical strength. A long post-curing step is challenging because it increases the overall processing time and, moreover, it may change the final dimensions of a printed part from the original settings. Accelerating the formation of the DA network can decrease the processing time and reduce the needs of a post-curing step. The gelation behavior of DA networks was studied by rheometry with respect to precursors of various architecture, functionality, and molecular weight. A precursor containing high functionality was prepared and examined for its impact on gelation behavior. A small fraction (~5 wt%) of highly functional microspheres brought in an efficient acceleration of the formation of DA networks. Next, DA epoxy samples were examined for the debonding-on-demand adhesive. Adhesive strength of DA epoxy samples was measured by lap shear testing as a function of temperature. Its adhesive strength maintained at the same level from room temperature to 120°C but dropped greatly (~20 times) above 120°C because of a transition by the retro-DA reaction. DA epoxy composite samples containing plasmonic nanoparticles were examined to present debonding under light exposure. This photothermal conversion can offer instant debonding exclusively at target area. Adhesive strength of re-attached samples was also measured and compared to those from pristine samples.

599a Overview of Caustic Aqueous Phase Electrochemical Reforming (CAPER) Technology for Process Intensified Hydrogen Production

Su Ha

The efficient generation of compressed pure hydrogen (H_2) is a critical milestone for sustaining H_2 in the energy network. Many of the traditional methods for compressed H_2 production are hindered by high operating temperatures, high energy costs, low efficiency, and complex unit operations. For example, the current water electrolysis technology requires a large energy of ~55 kWh to produce 1 kg of H_2 due to its high cell operating voltage of 1.9 V. Furthermore, the cost of liquifying, transporting, and storing H_2 from a large centralized source is approximately $14/kg, which must be significantly reduced to be market competitive. To address this H_2 production and distribution challenges, our Caustic Aqueous Phase Ethanol Reforming (CAPER) technology offers a novel electrochemical reforming reaction requiring a low thermoneutral cell voltage of 0.3V along with an *in-situ* CO_2 separation and its conversion to value-added product scheme enabling affordable, efficient, reliable, and secure either carbon neutral or carbon negative H_2 production for various fuel cell applications. The CAPER process reduces H_2O at the cathode to produce a high-pressure and high-purity H_2 at ≤80 °C while it oxidizes the ethanol at the anode using renewable electrical energy. The CO_2 produced by ethanol oxidation is also captured within the caustic electrolyte solution. The electrolyte can be refreshed to release compressed and pure CO_2 in a regenerator unit. Without needing downstream separation, compression, or membranes, the CAPER process achieves process-intensified H_2 production at high efficiency. For example, our CAPER flow reactor produces pure H_2 at 80 bar and 80 °C applying voltages at ≤ 0.5 V. The only gas-phase product detected was H_2 and any carbonaceous products were contained in the liquid phase. The H_2 was produced with 100% faradaic efficiency. The compressed H_2 was produced at ≤ 20 kWh kg^{-1}, which is much lower than electrolysis systems that require upwards of ~55 kWh kg^{-1}.

599b Mechanically Stable Polymer Molecular Sieve Membranes with Switchable Functionality

Tae-Hyun Bae

The creation of high-performance membranes that specifically target CO_2 is crucial for improving energy-efficient CO_2 capture methods. Although molecular sieves have been considered promising materials for membranes, their practical application has proven difficult. In this work, we present a novel method for developing a polymeric molecular sieve membrane that delivers superior CO_2 separation capabilities and maintains mechanical integrity. Initially, a polymer molecular sieve membrane with high gas permeability and mechanical stability was created using a specially formulated polymer that can be processed in solution, is highly cross-linkable, and can be functionalized. The selectivity for CO_2 was then enhanced by incorporating various amine-based compounds. Of these, polyethyleneimine was particularly effective, enhancing functionality in the larger pore areas while keeping the ultra-micropores intact, thus improving CO_2/N_2 separation efficiency. The resulting membrane shows exceptional CO_2/N_2 separation qualities, surpassing other polymer molecular sieve membranes and is competitive with many of the carbon molecular sieve membranes currently available.

599c Modeling Context Dependency in Interspecies Interactions: No Permanent Friends or Enemies in the Microbial Jungle

Hyun-Seob Song

Microorganisms are social – arguably more so than humans. In natural habitats, microorganisms form communities where they build diverse relationships with other species. Intriguingly, the networks of interacting species are not static; they continuously reorganize in response to environmental and compositional changes over space and time. Developing a predictive understanding of how these interactions vary across different contexts is crucial for microbiome engineering. This goal can be effectively achieved through the synergistic integration of data-driven and metabolic modeling approaches. In this talk, I will present recent advancements in computational and modeling methods that enable the prediction of how membership and environmental factors modulate microbial interactions.

599d Advancements in Electrochemical Methane Conversion: A Pathway to Enhanced Efficiency and Sustainability

Jun Hyuk Moon

The efficient conversion of methane into alcohols remains a pressing challenge with significant implications for resource optimization and environmental sustainability. Traditional, energy-intensive synthesis routes impose considerable limitations. Recently, electrochemical conversion has gained attention as a promising alternative due to its capability for direct partial oxidation of methane even under ambient temperature conditions. Here, we present our recent findings aimed at resolving these issues in electrochemical methane conversion. First, we introduce a methane-to-ethanol electrochemical conversion approach using an ionic oxidant, demonstrating a viable reaction energy pathway through combined experimental and computational methods. Second, we explain how a nanorod-structured catalyst enhances electrochemical methane conversion efficiency. Third, we present a novel concept of electrochemical methane conversion coupled with water splitting. Lastly, we discuss our investigation of a scalable and continuous reaction system.

599e Kiche President Young Investigator Award – Lithium Battery Recycling: Overview and a New Direction (Sponsored by SK Innovation)

Tae-Sik Oh

E ective recycling of lithium-ion battery waste can mitigate high-demand element supply chain problems, greenhouse gas emissions, and environmental hazards associated with the waste. For the cost-e ectiveness of the recycling process, the metal separation step needs to be as simplified as possible. Lithium, an alkaline metal, differs significantly in terms of chemistry from transition metals such as nickel, manganese, and cobalt; therefore, its separation from cathode waste is straightforward. Separation of nickel, manganese, and cobalt from each other is possible; however, it requires energy and time to recover each element with high purity and yield. This talk will review the recent trend in lithium battery recycling and present a new direction explored in my lab at Auburn University.

599f James Lee Memorial Award - a Fundamental Understanding of the Relationship between Molecular/Electronic Structure and Catalytic Performance (Sponsored by Hanwha Total)

Tae Jin Kim

Supported metal or metal oxide catalysts have been applied in diverse fields such as petrochemical industries, toxic gas remediation, pharmaceutical, the food industry, and more. It has been well documented that catalytic performance is controlled by the catalyst's composition, size, shape, and other factors. Along with the discovery of new catalyst synthesis methods, advanced catalyst characterization techniques have been extensively applied in catalysis. Understanding a molecular and electronic structure of fresh, used, deactivated, and reactive catalysts is crucial for the rational modification of currently used catalysts and the design of novel ones. A wide range of microscopic and spectroscopic techniques have been used to monitor the electronic structure, molecular structures, and interactions between surface species and support of heterogeneous catalysts. In addition to various characterization techniques, different characterization methodologies such as ex-situ, in-situ, and operando have been applied. These methodologies help elucidate the nature of active sites, provide the structure-activity-selectivity-stability relationship under working conditions (e.g., high temperature and pressure), and suggest possible reaction mechanisms based on the identification of intermediates. In this talk, I will provide an overview of catalyst structure-activity performance relationship, with a focus on ceria supported MOx (M = transition metal) catalysts. I will also present our recent results on a new catalyst synthesis method, one-pot chemical vapor deposition (OP-CVD) method, and its application in catalytic reaction such as reverse water gas shift reaction (RWGS) and CO oxidation. In the final section, I will highlight the importance of ex-situ and in-situ characterization methods for studying catalyst structures and revealing the molecular structure changes under the high and low temperatures.

303a Engineering Biology to Develop "a Taste for Waste"

Hal Alper

Chemical sustainability requires a fully circular economy. The valorization and remediation of waste (whether anthropogenic, environmental, or industrial) using biology presents itself as a viable route for both environmental stewardship and sustainable industrial bioproduction. This approach is becoming more feasible thanks to advances in metabolic strain engineering, synthetic biology, and machine learning. This talk will cover examples of how microbial systems and enzyme can achieve the goal of sustainable chemical production and waste valorization by rewiring biology to develop "a taste for waste". Highlighted examples include plastic waste depolymerization, unconventional carbon utilization, and feedstock expansions to include unique waste sources. Underlying metabolic and protein engineering principles will be highlighted through these examples. Collectively, this work helps to usher in a new approach for biologically-enabled sustainability.

249a My Journey in Chemical Engineering: from Initial Struggles to Development of Drug Delivery Systems and Tissue Engineering and Having Fantastic Students such as Cato Laurencin

Robert Langer

When I graduated from MIT, I went to work at Boston's Children's Hospital. I was the only engineer there. I began trying to see if I could make tiny particles that could deliver large molecules (e.g. proteins, nucleic acids) that prevent vascular growth to tumors. Over two years, I had found hundreds of unsuccessful methods. Finally, in 1976 I discovered a way to make it work (ultimately published in NATURE).

One of my goals was to move beyond just conducting experiments and publishing the results, to applying that work to helping people. We used the delivery systems to see if we could find substances that could stop blood vessels from growing. I isolated many different substances and tested them in over 100 different studies. We published a paper in Science in 1976, which showed for the first time that inhibitors of blood vessel inhibitors existed, and provided bioassays which could be used and were used to, isolate future inhibitors. Today, many inhibitors of blood vessel growth have been isolated and they are used by many millions of people to treat cancer and diseases of blindness, such as macular degeneration and diabetic retinopathy.

The principles we established for the controlled movement of molecules have been essential to the development of numerous clinically used therapeutics. There are many tiny controlled-release systems used by patients worldwide that continuously release peptides for up to six months from a single injection (Lupron Depot, Zoladex, and Decapeptyl) to treat advanced prostate cancer. Similar microspheres or other polymer systems containing bioactive molecules have led to new treatments for schizophrenia (Risperdal Consta), alcoholism, opioid addiction (Vivitrol), arthritis (Zilretta), controlling bleeding (Floseal, Surgiflo), pituitary dwarfism (Nutropin Depot), type-2 diabetes (Bydureon), heart disease (drug eluting stents) and many other diseases.

The original controlled-release materials we developed were small particles; in many cases, microparticles. However, nanoparticles are often critical for delivering significant payloads of any drug into cells, particularly newer potential drugs such as siRNA and mRNA. Yet, once nanoparticles are injected into the body, they are destroyed almost immediately by macrophages, and are unstable and often aggregate. These characteristics made their use essentially nonexistent. To address these issues, in a 1994 paper in Science, we defined seven characteristics we wanted to build into nanoparticles. We found that nanoparticles composed of a block copolymer of polyethylene glycol (PEG) and any other material and an added drug, could circulate for hours in vivo, be stable on the shelf for years, and not aggregate. Another issue with nanoparticles is that for nucleic acids, it is desirable that they be cationic (positively charged) so they can complex negatively charged nucleic acids. However, charged nanoparticles can cause toxicity. So, Dan Pack, David Putnam, and I added what are called ionizable molecules to nanoparticles that made them neutral at physiologic pH but charged inside the cells. This approach enables endosomal escape inside cells. Another area I worked on, initially with Jay Vacanti, was tissue engineering where Jay and I and my students created many of the principles and techniques in this field that are used today.

I also got involved in helping to start companies to advance our discoveries. The most well known company is Moderna. The company was initially criticized by scientists, stock analysts, and the news media who said the stock was overvalued and that our messenger RNA therapeutics would not work.

However, Moderna produced a COVID-19 vaccine that saved millions of lives. Many other new medical treatments for cancer and other diseases are in late-stage clinical trials. Polaris Ventures and Harvard Business Review stated that billions of people are or will be helped by products created by companies I helped start.

I'm also so proud of the students and postdocs who worked in our laboratory. When I turned 70, they had a celebration and over 700 people came. One of the students I'm so proud of is Cato Laurencin. He joined our laboratory in the 1980's and he has had a magnificent career and is a great person not only is he an outstanding scientist and mentor, he is also my friend. I'm so very honored to receive an award in his name.

105a Sustainability Efforts in Existing Chemical Process Facilities

Juan Inga

Most of the chemical industry was built 50 years ago. In the US there has not been a new large refinery built in the last 30 years. The chemical engineers that designed and built those facilities did it to meet the regulations placed at that time. Many facilities are now grandfathered in and therefore under no pressure to modify the existing facilities. Social pressure for change is being felt as the new workforce take over positions held before by Baby boomers. The generation called Gen-Z (now reaching 20) are different from previous generations since they are mostly driven by purpose instead of financial gain. Their new set of values and lifestyle are creating pressure in the chemical industry for products that have a reduced impact on the environment. This pressure will be expressed in our industry as new regulation or by consumer product selection. Should the chemical industry be proactive and work towards ways to reduce their carbon footprint and ensure a better position when this inevitable social pressure becomes the new normal. A few examples are discussed from the use of electricity to reduce fuel burning, to the reduction of flaring of associated gas.

508a Optimal Synthesis and Planning of SustainableChemical Process and Energy Systems

Ignacio Grossmann

In this lecture we give an overview on the applications of recent models and algorithms for the discrete and continuous optimization of a variety of challenging applications in the optimal synthesis and planning of sustainable chemical processes and energy systems. We provide an overview of applications of deterministic models based on mixed-integer linear/nonlinear programming (MILP/MINLP), Generalized Disjunctive Programming (GDP) and global optimization techniques, many of which have been developed by chemical engineers in the area of process systems engineering. We first consider applications of MILP in energy systems that include optimization of hydrogen supply chains, and the long term expansion planning of reliable and resilient electric power systems with high penetration of renewables. We also briefly discuss the retrofit of crude oil refineries for processing biomass to produce sustainable aviation fuels. Next, we consider applications of MINLP and GDP that include synthesis of lignocellulosic ethanol, optimization of shale gas infrastructures, and synthesis of intensified distillation columns. We consider next recent algorithms for rigorous global optimization of nonconvex optimization models for which we consider applications in optimal process water networks that involve reuse and recycle, optimal maritime transportation planning under carbon intensity constraints, and optimal design of centralized and distributed manufacturing facilities for biomass production. Finally, we briefly discuss extensions of these models to handle uncertainties. The variety of applications that we cover in this lecture, illustrates the scope that optimization techniques have towards the optimal synthesis and planning of sustainable chemical processes and energy systems.

621a Organizing Universities in the Century of Convergent Research

Robert A. Brown

The concept of the American research university was born in the second half of the 20th century with the government's acknowledgement of the roles of science and technology in defense and the growing demand for higher education. Their support expanded in the 1970's to encompass human health. Chemical engineering flourished as a discipline during the age of research universities, and I am a product of this success, spending my career in several of the very best departments. I've served as a teaching assistant for John McKetta at the University of Texas, took classes with Neal Amundson and Gus Aris at Minnesota, and was mentored by Skip Scriven there. I have experienced and led the work of great academic departments, which complicates what I will say in my lecture.

It is a special honor to make these remarks in a lecture named in Bill Schowalter's honor, one of the truly great chemical engineers of this era. Bill had an important influence on my career, as I will describe in my lecture.

During almost forty years as department head, dean, provost, and university president, I have thought a lot about the organization of research universities and its effect on their ability to address many of the most pressing problems facing mankind and to adapt to changes in our society. Performing research with this impact and educating the next generation of leaders who drive the needed changes are fundamental to the case for continuing support from the public and the government. Since the late 1800's we have relied on academic departments, formed around disciplines, to be the roots of university organization. The chemical engineering departments that I have been associated with — Texas, Minnesota, and MIT — are excellent examples, each doing well the three things that sociologists say they must; 1) be the academic home for students and faculty (the home of the tribe), 2) curate a body of knowledge that defines the discipline, while simultaneously serving as gatekeepers, and 3) anchor the labor market for doctoral graduates. Not surprisingly, the departments also are the advocacy groups within their institutions for resources, with the great ones being the most persuasive.

Departmental roles were traditionally performed independently, as disciplines were distinct, and, when the need arose for collaboration, universities created interdisciplinary centers and institutes to lure departmental faculty to work together. In this organizational structure, the departments — who hire and tenure faculty, and administer undergraduate and doctoral programs — are connected to interdisciplinary centers through their faculty members' participation.

The question I will address is whether this classical organization is the way forward in a world where the boundaries between disciplines — especially in STEM — are more diffuse than ever and the challenges we face, whether decarbonizing economies or harnessing AI for the good of society, require new levels of creativity and collaboration. Today, there is a call for fostering "convergent research" that requires much deeper interconnections between disciplines than traditionally fostered through interdisciplinary centers.

I believe that our historical academic structures are not tuned to optimize and sustain the needed levels of collaboration or to put the interconnections on the front burner of their planning and hiring. For the most part, departments still hire and tenure in relative isolation, making faculty allegiance to the academic tribe an essential commitment, an environment where convergent research collaborations are more difficult. I am concerned that present and future generations of young engineering scientists are less likely to want only these traditional affiliations and that the demand for talent also will drive rethinking our organization.

In my presentation I will describe two organizational innovations from my time at Boston University that address organizational change to drive convergent research. Both are aimed at recognizing the convergence of disciplines and the need to hire faculty who form the connections. In the first example, the College of Engineering, under the leadership of Dean Kenneth Lutchen, reorganized the College into three large departments and two divisions. Most importantly, after an extensive planning process, the faculty agreed on five convergent research themes. New faculty are hired *both* on the basis of perceived disciplinary needs and on finding talent for the convergent themes. Special search committees with representatives from all the departments and appropriate research centers (and in some cases outside of engineering) lead the search for convergent faculty members who will have faculty homes in the departments and centers. Over the last four years a quarter of new hires have been made by this search mechanism. Imagine a college of a 150 faculty members eventually with 40 - 50 faculty explicitly hired to support convergent research areas, bringing approaches from different disciplines to their teaching and research as their joint appointments interconnect academic departments further blurring the knowledge boundaries. The result will give us the needed scale to make impact in the convergent themes and create a novel environment for collaborative research and education.

The infusion of machine learning and AI across all of a university represents a much larger challenge requiring hiring faculty with deep expertise in rapidly evolving data science and computing who, in turn, will these methods to all the areas of application. To do this we established the Faculty and Computing & Data Sciences (CDS) as an all-university academic unit composed of a core faculty and other members with multiple departmental affiliations. The community of like-minded data scientists formed by CDS gives the initiative a scale that would be impossible with fragmented traditional academic hiring and prevents the inertia caused by embedding it in a traditional computer science department.

As I will emphasize in the lecture, the idea is not to destroy academic disciplines; they are and will continue to be the foundation of academic organization. Our goal is to make the boundaries more diffuse and permeable allowing for the needed interconnections as domains of knowledge and application change and evolve.

Finally, my experience has reenforced the reality that changing academic organization is perhaps the most difficult role of university leadership and can only be accomplished when the management and financial systems are strategically centered as well. Which is the more difficult task is an open debate. The difficulty of both challenges is not new: already in 1532 Machiavelli wrote:

"It must be considered that nothing is more difficult to carry out, nor more doubtful of success, nor more dangerous to handle, than to initiate a new order of things. For the reformer has enemies in all those who profit by the old order, and only lukewarm defenders in all those who profit by the new order."[1]

Organization change is needed to be able to continually realign research universities with the societal demands on higher education. The question is whether we have the energy and will to move in that direction.

[1] Machiavelli, N. T. (2010). *The Prince.* University of Chicago Press, Chicago. Originally published in 1532.

738a Molecular Layer Deposition of Polyhydroquinone Thin Films for Li Ion Battery Applications

Matthias Young

Many next-generation materials for Li-ion batteries are limited by material instabilities. To stabilize these materials, ultrathin, protective coatings are needed that conduct both lithium ions and electrons. Here, we demonstrate a hybrid chemistry combining molecular layer deposition (MLD) of trimethylaluminum (TMA) and p-hydroquinone (HQ) with oxidative molecular layer deposition (oMLD) of molybdenum pentachloride ($MoCl_5$) and HQ to enable vapor-phase molecular layer growth of poly(p-hydroquinone) (PHQ) – a mixed electron and lithium ion conducting polymer. We employ quartz crystal microbalance (QCM) studies to understand the chemical mechanism and demonstrate controlled linear growth with a 0.5 nm/cycle growth rate. Spectroscopic characterization indicates that this hybrid MLD/oMLD chemistry polymerizes surface HQ monomers from the TMA-HQ chemistry to produce PHQ. The polymerization to PHQ improves air stability over MLD TMA-HQ films without crosslinking. Electrochemical measurements on hybrid MLD/oMLD films indicate electronic conductivity of $\sim 10^{-9}$ S/cm and a Li-ion conductivity of $\sim 10^{-4}$ S/cm. While these coatings show promise for Li-ion battery applications, this work focuses on establishing the coating chemistry and future studies are needed to examine the stability, structure, and cycling performance of these coatings in full Li-ion cells.

738b Towards Understanding the Role of Transport in Electrocatalysis: the Role of Dimensionless Analysis and Multi-Scale Modeling

Carlos Morales-Guio, Joonbaek Jang, Momoko Ishida

promising approach to store energy is the use of excess electricity to drive the production of fuels and chemicals in electrochemical reactors. The lack of understanding of the relation between transport phenomena (e.g. mass, heat, charge), intrinsic kinetics and electrocatalytic performance is currently limiting the scale-up of electrocatalytic technologies particularly for transformations where transport phenomena play a role in determining product selectivity and production rate. In this presentation, I will discuss our approach of combining reactors design, multi-scale modeling and dimensionless analysis to understand and decouple mass, charge and heat transfer effects from intrinsic electrode kinetics particularly for transformations that involve dissolved gas substrates (i.e. H_2, O_2, CO_2, CO, CH_4, N_2) and diluted protons in near neutral pH.

In this talk, the development of a multi-scale first-principles reaction-transport model is presented for the electrochemical reduction of CO_2 to fuels and chemicals on polycrystalline copper electrodes. The model utilizes a continuous stirred-tank reactor (CSTR)-volume approximation that captures the relative timescales for mesoscale stochastic processes at the electrode/electrolyte interface that determine product selectivity. The model is built starting from a large experimental dataset obtained under a broad range of well-defined transport regimes in a gastight rotating cylinder electrode cell. Product distributions under different conditions of transport, applied potential, bulk electrolyte concentration, and catalyst porosity are rationalized by introducing dimensionless numbers that reduce complexity and capture relative timescales for mesoscopic and microscopic dynamics of electrocatalytic reactions on copper electrodes of any porosity. This work demonstrates that one CO_2 reduction mechanism can explain differences in selectivity reported for copper-based electrocatalysts when mass transport, concentration polarization effects, and primary and secondary current distributions are taken into account. The reaction-transport model presented here should enable the rational design of CO_2 electrolyzers.

738c Multiscale Simulation Framework to Study Plastic Transformations

Sophia Ezendu, Ademola Soyemi, Tibor Szilvasi

Plastic transformations are critical to ongoing recycling and upcycling efforts, but the complexity of the reactions makes it difficult to understand the effect of individual factors on reaction rates and product distributions experimentally. In this work, we report on a multiscale simulation framework for studying polymer transformations that incorporates affordable high-level coupled cluster calculations combined with benchmarked density functional theory calculations, detailed conformer search, and lattice-based kinetic Monte Carlo simulations to provide the temporal and spatial evolution of the polymer during transformations. Our framework can match experimentally observed reaction times within an order of magnitude without any parameter estimation in base-assisted dehydrochlorination of polyvinyl chloride. We determine that the E2 reaction mechanism dominates the reaction and demonstrate that different structural defects can inhibit or promote directional polyene growth as well as affect the structure of the dehydrochlorination product.

Reference: S. Ezendu, A. Soyemi, T, Szilvási, *AIChE Journal*, 70, e18559 (2024).

738d Insights into Intrinsic Kinetics for HDPE Pyrolysis through Experiments and Data Science

J.V. Jayarama Krishna, Barbara Alejandra Perez, Hilal Ezgi Toraman

Chemical recycling, particularly through pyrolysis, has emerged as a promising solution for plastic waste management. It has the potential to process both single and mixed plastics, transforming them into value-added chemicals, monomers, and fuels. Polymer pyrolysis results in complex reaction networks involving thousands of species and reactions. Factors such as temperature, residence time, heating rate, and sample characteristics influence product distribution and play vital roles in kinetic measurements as well as the extent of secondary reactions. Polymer samples, particularly those with large length scales, such as packed powders and pellets, can significantly impact kinetics due to heat and mass transfer rates within the polymer sample. This underscores the importance of selecting the appropriate reactor configuration and establishing well-defined experimental conditions to minimize the transport limitations.

Using a combination of well-defined and well-controlled experimentation, data science, and advanced analytics, this work methodically studied the effect of various pyrolysis variables on the product distribution of high-density polyethylene (HDPE) to validate the pyrolysis conditions that minimize transport effects and secondary gas-phase reactions. The primary decomposition of HDPE was performed using a Box–Behnken design to evaluate the role of pyrolysis variables such as particle size, sample size, carrier gas flow rate, and temperature. The pyrolysis of HDPE under various experimental conditions resulted in the identification of nearly 500 products using GC X GC/TOF–MS and quantification with FID. The chemical groups identified for the HDPE pyrolysis included paraffins, olefins, diolefins, and cyclic and aromatic hydrocarbons among others. The film

thickness calculated using characteristic time analysis for the sample sizes (50–150 μg) employed in this work varied in the range of 4.7–14.1 μm. The analysis indicated that the experiments conducted within the BBD space were well within the isothermal, reaction-controlled regime. Principal component analysis (PCA) demonstrated statistical differences for temperature (480–600 C), while no statistical differences were observed for particle size (53–125 to >300 μm), sample size (50–150 μg), and carrier gas flow

rate (100–300 mL min−1). The total yield of C3 to C28 increased from 25.98 1.28 to 55 1.42 wt % when temperature was increased from 480 to 600 C. When the sample size of 1000 μg was considered for pyrolysis, PCA analysis showed statistical differences in the product distribution with respect to the smaller sample sizes of 50– 150 μg, which can possibly indicate the presence of transport effects at such higher sample sizes. This work underscores the importance of carefully selecting the pyrolysis parameters to curtail secondary gas-phase reactions and minimize transport effects. The product speciation data obtained under such experimental conditions are valuable for developing microkinetic models for polymer pyrolysis.

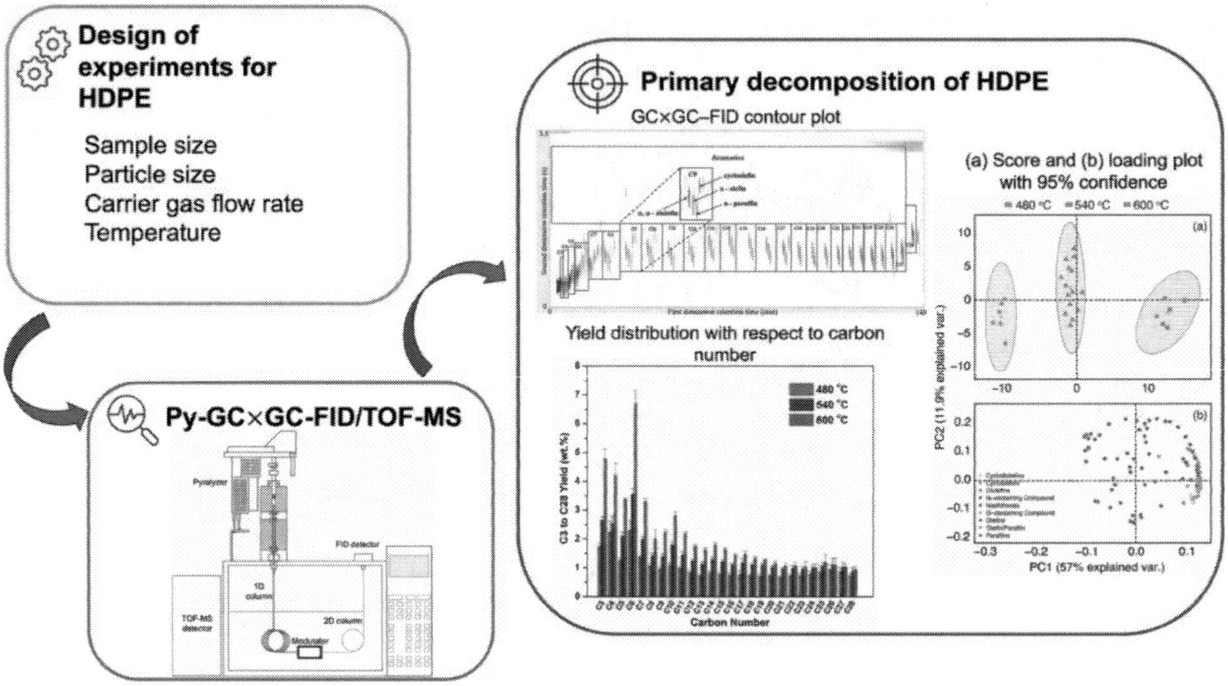
Design of experiments for HDPE
Sample size
Particle size
Carrier gas flow rate
Temperature
Py-GC×GC-FID/TOF-MS
Pyrolyzer
FID detector
1D column
2D column
TOF-MS detector
Modulator
Primary decomposition of HDPE
GC×GC–FID contour plot
Yield distribution with respect to carbon number
480 °C
540 °C
600 °C
C3 to C28 Yield (wt.%)
Carbon Number
(a) Score and (b) loading plot with 95% confidence
PC2 (11.9% explained var.)
PC1 (57% explained var.)

738e Thermodynamic Implications of Size, Hydrophilicity, and Fluorine Content on Perfluoroalkyl Adsorption in NU-1000

Phillip Johnson, Brandon C. Bukowski

The molecular mechanisms of adsorption are central to engineering new adsorbents to capture environmental contaminants, such as per- and polyfluoroalkyl substances (PFAS). Metal-organic frameworks (MOFs) have been shown to adsorb some PFAS classes, yet a fundamental understanding of how PFAS identity in the presence of water affects their adsorption capacity is not known. We performed grand canonical Monte Carlo simulations and molecular dynamics of perfluoroalkanoic acids (PFAAs) adsorption in MOF NU-1000 under aqueous conditions with varying carbon chain length to interrogate how PFAS chemical structure affects adsorption capacity. We found that larger PFAAs adsorb more favorably into NU-1000 than smaller chain PFAAs due to the formation of aggregates in the mesopores that stabilize anionic adsorption to the node. Due to their size and hydrophilicity, smaller chains tend to limit interactions with the adsorbent. We characterized aggregate formation based on the strength of node and linker interactions. These insights offer directions to develop novel materials to promote aggregate formation to capture and retain a wider set of PFAS from aqueous solutions. We then describe post-synthetic strategies to promote aggregate formation where nonpolar interactions may improve short chain perfluoroalkane adsorption.

738f Active Learning for Molecular Simulations of Adsorption in Metal-Organic Frameworks

Yamil Colón

Metal-organic frameworks (MOFs) are porous, crystalline materials comprised of inorganic nodes and organic linkers. Their porous nature along with the potential to tailor the chemistry of the pores, makes them very attractive for various adsorption applications. However, the sheer number of combinations of building blocks leads to limitless possible structures, raising questions as to the feasibility of characterizing them across the thermodynamic conditions needed to evaluate them in various applications. Herein I will present our efforts to accurately and efficiently determine adsorption in databases of MOFs using active learning (AL) techniques combined with molecular simulations of adsorption. We use AL to sequentially determine the next molecular simulation to be performed while simultaneously developing a surrogate model capable of describing adsorption in the feature space selected. We first established and validated the AL technique while varying temperature, pressure, and composition of the adsorbates, finding that AL can accurately describe the adsorption space with significantly less data. We then expand the analysis to include the MOF space and adsorbate space within the AL framework and find significant data savings. Lastly, we explore the use of reinforcement learning (RL) for adsorption AL campaigns in MOF databases.

738g Accelerating Process Development in Biomanufacturing via Digitalisation

Maria M Papathanasiou, Foteini Michalopoulou, Steven Sachio, Andrea Galeazzi

The Life Sciences sector, including pharmaceuticals and medtech, is at the forefront of the global economy, accounting for a total value of 285 billion U.S. dollars in 2020 (Mikulic, 2024). When it comes to process development and innovation, the sector's leading position is not reflected in the day-to-day operation. Manufacturers develop processes by relying on time- and cost- intensive wet-lab experimentation. This increases the use of solvents and raw materials, negatively impacting the environmental footprint, while delaying the time-to-market of novel therapeutics (Kasemiire et al., 2021). For this, systematic approaches are required that can offer a cost- and time- efficient platforms to screen materials and operating conditions sustainably. In this context, digital tools offer a cost- and time- effective solution to perform *in silico* experiments that investigate the system dynamics and assist with decision-making during process development.

Model-based approaches are dependent on the existence of a reliable digital twin that describes the process dynamics. This can be either a mechanistic model that is based on first principles, a black-box model derived directly from data or a hybrid formulation. Each model structure requires diverse types of data and dataset sizes to be available for development and validation (Michalopoulou & Papathanasiou, 2024). Irrespective of the model structure, however, biopharmaceutical systems are often challenged by complex dynamics that are poorly understood and further impaired by unavailability of measurements suitable for model development. The choice of modelling approach becomes therefore conditional to the specific requirements and capabilities of the unit operation and/or system at hand. As a result, a model-based identification of a suitable operating region (design space) should be tailorable to the model type and the system specifications (Sachio et al., 2023; Sachio et al., 2024).

In this talk, present a portfolio of digital tools for modelling, optimisation and design space identification in the area of biopharmaceutical manufacturing. We focus on biomanufacturing separation systems and we assess different modelling approaches and their capabilities in serving as reliable digital twins. We further present a computer-modelling framework for design space identification that caters for both systems for which a mechanistic model is available and for occasions where fully data-driven approaches are employed. Lastly, we discuss how our tools are applicable to other sectors within Chemical Engineering, such as energy systems.

References

Kasemiire, A., Avohou, H. T., De Bleye, C., Sacre, P. Y., Dumont, E., Hubert, P. & Ziemons, E. 2021. Design of experiments and design space approaches in the pharmaceutical bioprocess optimization. *European Journal of Pharmaceutics and Biopharmaceutics*, 166, 144-154.

Michalopoulou, F., & Papathanasiou M. M. 2024. Assessment of data-driven modeling approaches for chromatographic separation processes. *AIChE Journal*, e18600

Mikulic, M. 2024. Global value of life sciences sector deals 2013-2021. *Statista*, [online] URL: https://www.statista.com/statistics/398216/total-value-of-global-deals-in-chemical-pharmaceutical-industry/

Sachio, S., Kontoravdi, C. & Papathanasiou, M. M. 2023. A model-based approach towards accelerated process development: A case study on chromatography. *Chemical Engineering Research and Design*, 197, 800-820.

Sachio, S., Likozar, B., Kontoravdi, C. & Papathanasiou, M. M. 2024. Computer-aided design space identification for screening of protein A affinity chromatography resins. *Journal of Chromatography A*, 1722, 464890.

AUTHOR INDEX

AIChE
120 Wall Street, FL 23
New York, NY 10005-4020

ISBN 979-8-3313-1688-4
9 798331 316884